1 水利固定资产投资

全年水利建设完成投资 6602.6 亿元，较上年减少 529.8 亿元，减少 7.4%。其中：建筑工程完成投资 4877.2 亿元，较上年减少 3.8%；安装工程完成投资 280.9 亿元，较上年增加 5.7%；机电设备及各类工器具购置完成投资 214.4 亿元，较上年增加 1.3%；其他完成投资（包括移民征地补偿等）1230.1 亿元，较上年减少 22.4%。

	2011年/亿元	2012年/亿元	2013年/亿元	2014年/亿元	2015年/亿元	2016年/亿元	2017年/亿元	2018年/亿元	2018年比上年增加比例/%
全年完成	3086.0	3964.2	3757.6	4083.1	5452.2	6099.6	7132.4	6602.6	-7.4
建筑工程	2103.2	2736.5	2782.8	3086.4	4150.8	4422.0	5069.7	4877.2	-3.8
安装工程	121.7	237.8	173.6	185.0	228.8	254.5	265.8	280.9	5.7
设备及各类工器具购置	115.2	178.1	161.1	206.1	198.7	172.8	211.7	214.4	1.3
其他（包括移民征地补偿等）	745.9	811.8	640.2	605.6	873.9	1250.3	1585.2	1230.1	-22.4

在全年完成投资中，防洪工程建设完成投资 2175.4 亿元，较上年减少 10.8%；水资源工程建设完成投资 2550.0 亿元，较上年减少 5.7%；水土保持及生态工程建设完成投资 741.4 亿元，较上年增加

8.6%;水电、机构能力建设等专项工程完成投资1135.8亿元,较上年减少13.0%。

七大江河流域完成投资5108.6亿元,东南诸河、西北诸河以及西南诸河等其他流域完成投资1494.0亿元;东部、中部、西部、东北地区完成投资分别为2335.3亿元、1638.8亿元、2417.7亿元和210.8亿元,占全部完成投资的比例分别为35.4%、24.8%、36.6%和3.2%。

在全年完成投资中,中央项目完成投资116.6亿元,地方项目完成投资6486.0亿元。大中型项目完成投资1183.5亿元,小型及其他项目完成投资5419.1亿元。各类新建工程完成投资5149.6亿元,扩建、改建等项目完成投资1453.0亿元。

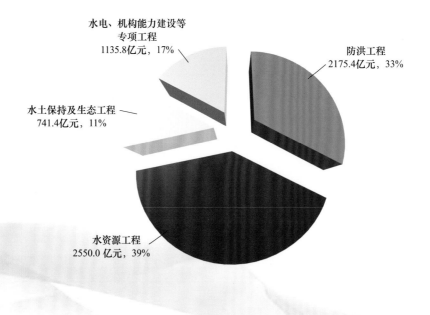

2018年分用途完成投资情况

2018年

全国水利发展统计公报

2018 Statistic Bulletin
on China Water Activities

中华人民共和国水利部 编

Ministry of Water Resources, People's Republic of China

·北京·

图书在版编目（CIP）数据

2018年全国水利发展统计公报 = 2018 Statistic Bulletin on China Water Activities / 中华人民共和国水利部编. -- 北京：中国水利水电出版社，2019.10
ISBN 978-7-5170-8101-2

Ⅰ. ①2… Ⅱ. ①中… Ⅲ. ①水利建设－经济发展－中国－2018 Ⅳ. ①F426.9

中国版本图书馆CIP数据核字(2019)第235814号

书　　名	2018 年全国水利发展统计公报 2018 Statistic Bulletin on China Water Activities 2018 NIAN QUANGUO SHUILI FAZHAN TONGJI GONGBAO
作　　者	中华人民共和国水利部　编 Ministry of Water Resources, People's Republic of China
出版发行	中国水利水电出版社 （北京市海淀区玉渊潭南路1号D座　100038） 网址：www.waterpub.com.cn E-mail：sales@waterpub.com.cn 电话：（010）68367658（营销中心）
经　　售	北京科水图书销售中心（零售） 电话：（010）88383994、63202643、68545874 全国各地新华书店和相关出版物销售网点
排　　版	中国水利水电出版社微机排版中心
印　　刷	北京博图彩色印刷有限公司
规　　格	210mm×297mm　16开本　3.75印张　52千字
版　　次	2019年10月第1版　2019年10月第1次印刷
印　　数	0001—1000 册
定　　价	**28.00 元**

凡购买我社图书，如有缺页、倒页、脱页的，本社营销中心负责调换

版权所有·侵权必究

目 录

1 水利固定资产投资 ………………………………………… 1
2 重点水利建设 ……………………………………………… 4
3 主要水利工程设施 ………………………………………… 7
4 水资源利用与保护 ………………………………………… 12
5 防洪抗旱 …………………………………………………… 14
6 水利改革与管理 …………………………………………… 16
7 水利行业状况 ……………………………………………… 23

Contents

I. Investment in Fixed Assets　27
II. Key Water Projects Construction　31
III. Key Water Facilities　34
IV. Water Resources Utilization and Protection　40
V. Flood Control and Drought Relief　41
VI. Water Management and Reform　44
VII. Current Status of Water Sector　51

  2018年是全面贯彻党的十九大精神的开局之年，也是水利事业承前启后的重要一年。根据中央关于机构改革的部署，国务院三峡办、国务院南水北调办并入水利部，水利部机构职能进行了优化调整，水利事业开启了新的征程。党中央、国务院高度重视水利工作。一年来，各级水利部门在习近平新时代中国特色社会主义思想指引下，认真贯彻党中央、国务院决策部署，真抓实干、迎难而上，推动各项水利工作取得新进展。

全年水利建设新增固定资产 3610.8 亿元。截至 2018 年年底，在建项目累计完成投资 16745.9 亿元，投资完成率为 63.0%；累计新增固定资产 9459.2 亿元，固定资产形成率为 56.5%，比上年增加 1.0 个百分点。

当年施工的水利建设项目 27930 个，在建项目投资总规模 27499.8 亿元，较上年增加 10.0%。其中：有中央投资的水利建设项目 16928 个，较上年增加 8.8%；在建投资规模 14204.7 亿元，较上年增加 7.1%。新开工项目 19786 个，较上年增加 0.3%，新增投资规模 6060.2 亿元，比上年减少 33.2%。全年水利建设完成土方、石方和混凝土方分别为 33.8 亿立方米、4.5 亿立方米、1.0 亿立方米。截至 2018 年年底，在建项目计划实物工程量完成率分别为：土方 78.2%、石方 67.7%、混凝土方 63.4%。

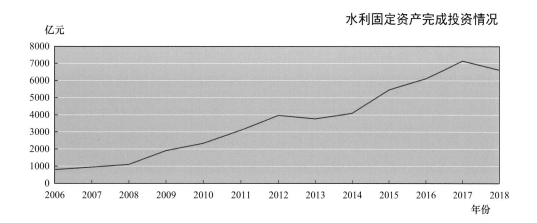

2 重点水利建设

江河湖泊治理。全年在建江河治理工程5406处，其中：堤防建设535处、大江大河及重要支流治理658处、中小河流治理3677处、行蓄洪区安全建设及其他项目536处。截至2018年年底，在建项目累计完成投资3386.3亿元，项目投资完成率64.9%。长江中下游河势控制和河道整治深入推进；黄河下游防洪工程加快实施；进一步治淮38项工程已开工29项，其中6项建成并发挥效益；东北三江治理基本完工；太湖流域水环境综合治理21项工程已开工18项，其中12项已建成并发挥效益。

水库及枢纽工程建设。全年在建水库及枢纽工程1033座，截至2018年年底，在建项目累计完成投资2942.6亿元，项目投资完成率58.1%。陕西东庄、湖北碾盘山、福建白濑、内蒙古东台子、安徽牛岭等大型水库及枢纽和秦镜水库、鱼枧水库等7座中型水库开工，双溪水库、黄沙水库等34座中型水库加快建设，黑河黄藏寺、福建霍口、贵州黄家湾、云南车马碧、广西驮英等大型水库及枢纽实现年度截流目标，辽宁猴山水库下闸蓄水。

水资源配置工程建设。全年水资源配置工程在建投资规模7327.8亿元，累计完成投资4095.0亿元，项目投资完成率55.9%。开展了71个河湖水系连通项目建设，改善了379余条（个）河流（湖泊或水库）的连通性。

农村水利建设。全年农村饮水安全巩固提升工程完成投资573.6亿元，其中中央补助资金76.3亿元；受益人口7800多万人，其中涉及国家建档立卡贫困人口436万人。截至2018年年底，农村自来水普及率达到81%，农村集中供水率达到86%。当年安排中央投资用于大中型灌排工程建设与灌区节水改造137.6亿元、高效节水灌溉等农田水利建设225亿元。全年新增耕地灌溉面积828千公顷，新增节水灌溉面积2022千公顷，新增高效节水灌溉面积1536千公顷。

农村水电建设。全年农村水电建设完成投资100.1亿元，新增水电站194座，投产发电设备容量164.3万千瓦，其中：新投产装机116.1万千瓦，技改净增发电设备容量48.2万千瓦。

水土流失治理。全年水土保持及生态工程在建投资规模886.3亿元，累计完成投资511.9亿元。全国新增水土流失综合治理面积6.4万平方公里，其中国家水土保持重点工程新增水土流失治理面积1.25万平方公里。对560座黄土高原淤地坝进行了除险加固。

行业能力建设。 全年水利行业能力建设完成投资45.3亿元，其中：防汛通信设施投资7.1亿元，水文建设投资13.1亿元，科研教育设施投资1.9亿元，其他投资23.2亿元。

3 主要水利工程设施

堤防和水闸。截至 2018 年年底，全国已建成 5 级及以上江河堤防 31.2 万公里[1]，累计达标堤防 21.8 万公里，达标率为 69.8%；其中 1 级、2 级达标堤防长度为 3.4 万公里，达标率为 80.5%。全国已建成江河堤防保护人口 6.3 亿人，保护耕地 4.1 万千公顷。全国已建成流量 5 立方米每秒及以上的水闸 104403 座，其中大型水闸 897 座。按水闸类型分，分洪闸 8373 座，排（退）水闸 18355 座，挡潮闸 5133 座，引水闸 14570 座，节制闸 57972 座。

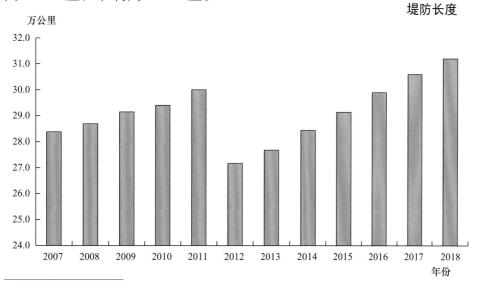

[1] 2011 年以前各年堤防长度含部分地区 5 级以下江河堤防长度。

水库和枢纽。全国已建成各类水库98822座，水库总库容8953亿立方米。其中：大型水库736座，总库容7117亿立方米，占全部总库容的79.5%；中型水库3954座，总库容1126亿立方米，占全部总库容的12.6%。

机电井和泵站。全国已建成日取水量大于等于20立方米的供水机电井或内径大于等于200毫米的灌溉机电井共510.1万眼。全国已建成各类装机流量1立方米每秒或装机功率50千瓦及以上的泵站95468处，其中：大型泵站376处，中型泵站4276处，小型泵站90816处。

灌区工程。全国已建成设计灌溉面积大于2000亩及以上的灌区共22873处，耕地灌溉面积37752千公顷。其中：50万亩以上灌区175处，耕地灌溉面积12399千公顷；30万～50万亩大型灌区286处，耕地灌溉面积5400千公顷。截至2018年年底，全国灌溉面积74542千公顷，其中耕地灌溉面积68272千公顷，占全国耕地面积的50.7%。全国节水灌溉工程面积36135千公顷，其中：喷灌、微灌面积11338千公顷，低压管灌面积10566千公顷。

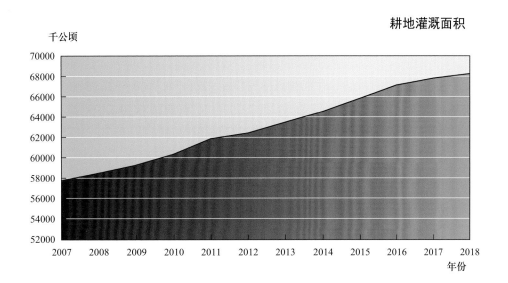

耕地灌溉面积

农村水电。全国已建成农村水电站46515座，装机容量8043.5万千瓦，占全国水电装机容量的22.8%；年发电量2345.6亿千瓦时，占全国水电发电量的19.0%。

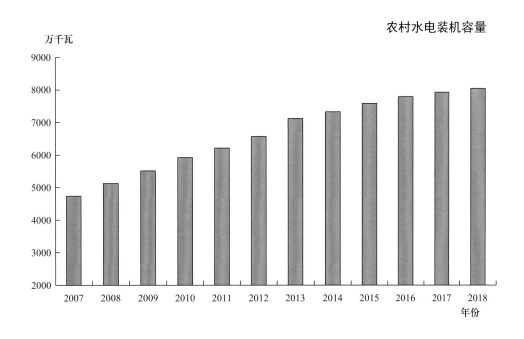

农村水电装机容量

水土保持工程。全国累计水土流失综合治理面积达 131.5 万平方公里❶，累计封禁治理保有面积达 23.43 万平方公里。2018 年水土流失动态监测工作中央和地方分级负责、协同开展，实现全国覆盖，监测面积约 960 万平方公里，全面掌握县级行政区及重点区域的水土流失动态变化。

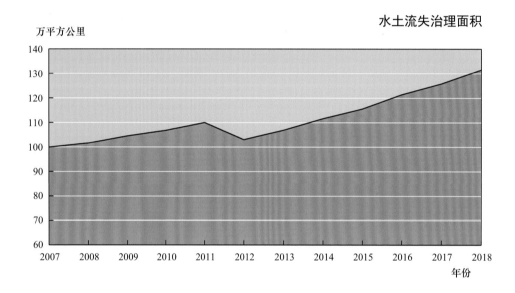

水文站网。全国已建成各类水文测站 121097 处，其中：国家基本水文站 3154 处，专用水文站 4099 处，水位站 13625 处，雨量站 55413 处，蒸发站 19 处，地下水站 26550 处，水质站 14286 处，墒情站 3908 处，实验站 43 处。向县级以上防汛指挥部门报送水文信息的各类水文测站 66439 处，可发布预报站 1887 处。配备在线测流系统的水文测站 1616 处。已基本建成由中央、流域、省级和地市级共 332 个监测机构

---

❶ 2012 年数据与第一次全国水利普查数据进行了衔接。

组成的覆盖全国主要江河湖库和重点地区地下水的水质监测体系。

水利网信。截至 2018 年年底，省级以上水利部门配置各类服务器 5764 台（套），形成存储能力 18PB，存储各类信息资源总量达 2.4PB；全国县级以上水利部门配置各类卫星设备 3895 台（套），具备北斗卫星短报文传输能力的报汛站达 6549 多个，配置应急通信车 48 辆、无人机 819 架、集群通信终端 5313 个；全国省级以上水利部门各类信息采集点达 45 万处，其中：水文、水资源、水土保持等采集点约 22.76 万个，大中型水库安全监测采集点约 22.5 万个。

4 水资源利用与保护

水资源状况。全年全国水资源总量27462.5亿立方米，与多年平均值基本持平；全国年平均降水量682.5毫米，比多年平均偏多6.2%，较上年增加2.7%。截至2018年年底，全国669座大型水库和3602座中型水库年末蓄水总量4104.3亿立方米，比年初减少38.0亿立方米。

水资源开发。全年新增规模以上水利工程❶供水能力101.6亿立方米。截至2018年年底，全国水利工程供水能力达8676.8亿立方米，其中：跨县级区域供水工程567.1亿立方米，水库工程2323.7亿立方米，河湖引水工程2105.1亿立方米，河湖泵站工程1754.7亿立方米，机电井工程1400.9亿立方米，塘坝窖池工程357.9亿立方米，非常规水资源利用工程168.4亿立方米。

---

❶ 规模以上水利工程包括：总库容大于等于10万立方米水库、装机流量大于等于1立方米每秒或装机容量大于等于50千瓦的河湖取水泵站、过闸流量大于等于1立方米每秒的河湖引水闸、井口井壁管内径大于等于200毫米的灌溉机电井和日供水量大于等于20立方米的机电井。

水资源利用。全年总供水量 6015.5 亿立方米，其中：地表水源占 82.3%，地下水源占 16.2%，其他水源占 1.5%。全国总用水量 6015.5 亿立方米，其中：生活用水 859.9 亿立方米，占总用水量的 14.3%；工业用水 1261.6 亿立方米，占总用水量的 21.0%；农业用水 3693.1 亿立方米，占总用水量的 61.4%；人工生态环境补水 200.9 亿立方米，占总用水量的 3.3%。与上年比较，用水量减少 27.9 亿立方米，其中：农业用水量减少 73.3 亿立方米，工业用水量减少 15.4 亿立方米，生活用水及人工生态环境补水量分别增加 21.8 亿立方米和 39.0 亿立方米。全国人均综合用水量为 432 立方米，农田灌溉水有效利用系数 0.554，万元国内生产总值（当年价）用水量 66.8 立方米，万元工业增加值（当年价）用水量 41.3 立方米。按可比价计算，万元国内生产总值用水量和万元工业增加值用水量分别比 2017 年下降 6.6% 和 6.9%。

河湖水质。根据对全国 26.2 万公里河流水质评价结果，河流水质达到Ⅰ～Ⅲ类的河长占 81.6%。

5 防洪抗旱

2018年，全国洪涝灾害总体偏轻，洪涝灾害直接经济损失占当年GDP的百分比为0.18%。全国农作物受灾面积6426.98千公顷，成灾面积3131.16千公顷，受灾人口0.56亿人，因灾死亡187人，失踪32人，倒塌房屋8.51万间，县级以上城市受淹83个，直接经济损失1615.47亿元，其中水利设施直接经济损失257.98亿元。四川、山东、广东、甘肃、内蒙古、云南等省（自治区）受灾较重。全国因山洪灾害造成人员死亡和失踪占全部死亡和失踪人数的73.52%，因台风造成经济损失占全国洪涝灾害直接经济损失的比例为41.71%。

全国旱灾总体偏轻，辽宁、内蒙古等省（自治区）旱灾比较严重。全国农田因旱受灾面积7379.21千公顷，成灾面积3667.23千公顷，直接经济总损失483.62亿元。全国因旱累计有306.69万城乡人口、462.30万头大牲畜发生临时性饮水困难。

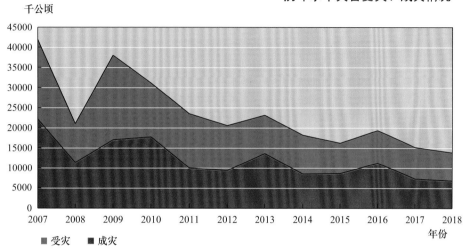

历年水旱灾害受灾、成灾情况

全年中央下拨特大防汛抗旱经费 35.06 亿元，其中：特大防汛经费 30.99 亿元，特大抗旱经费 4.06 亿元。全年防洪减淹耕地 3176.08 千公顷，避免城市受淹 75 座次，防洪减灾经济效益 239.98 亿元。解决了 476.98 万城乡居民和 244.24 万头大牲畜因旱临时饮水困难，全年完成抗旱浇地面积 23422.06 千公顷，挽回粮食损失 221.38 亿公斤。各地防汛抢险累计投入 467.95 万人次，出动舟船 1.78 万舟次、运输设备 20.20 万班次、机械设备 14.68 万班次，消耗物资价值 13.15 亿元，紧急组织转移群众 807.31 万人次，开动机电井 467.87 万眼、泵站 3.66 万处、机动抗旱设备 568.59 万台（套），出动各类运水车 93.65 万辆次。

6 水利改革与管理

河（湖）长制。截至 2018 年年底，全国全面建立河长制湖长制，共明确省、市、县、乡四级河长、湖长 30 多万名，各地设立村级河长、湖长（含巡河员、护河员）90 多万名。各省（自治区、直辖市）均设置省、市、县级河长制办公室，建立了配套制度，党政负责、水利牵头、部门联动、社会参与的工作格局基本形成。

最严格水资源管理。水利部会同国家发展改革委等完成对 31 个省（自治区、直辖市）2017 年度最严格水资源管理制度考核；江苏、山东、安徽、重庆、北京、浙江、上海 7 个省（直辖市）考核等级为优秀。太湖、松花江干流等 13 条跨省江河流域水量分配方案得以批复，河北、辽宁等 13 个省（区）开展了省内跨市县江河水量分配工作。2017—2018 调水年度南水北调东线一期工程向山东调水 10.88 亿立方米，中线一期工程向北京、天津、河北、河南 4 省（直辖市）调水共计 74.58 亿立方米，受水区供水安全保障能力显著提升。加强黄河、黑河等重要河流水资源统一调度，黄河干流实现连续 19 年不断流，黑河下游东居延海连续 14 年不干涸。以县域为单元全面开展节水型社会达标建设，北京等8个省（自治区、直辖市）共计65个县（区）级行政区通过达标验收。105个水生态文明试点城市建设进展顺利，其中96个试点城市通过验收。实施华北地下水超采综合治理河湖地下水回补试点工作，利用南水北调中线工程和河北省水库对河北省3条河流进行地下水回补，累计补水6.27亿立方米。南水北调东中线一期工程受水区累计压采地下水19亿立方米。宁夏、内蒙古、广东、河南、甘肃、江西、湖北等7个水权试点全部完成验收。中国水权交易所2018年完成水权交易51单，成交水量13.31亿立方米，自开业以来累计完成水权交易92单，成交水量27.74亿立方米。

运行管理。截至2018年年底，全国纳入水管体制改革范围的水管单位14325个，经精简撤并调整为12908个，较改革前下降10%。共落实两项经费301.4亿元，落实率达88.1%，其中：落实公益性人员基本支出177.7亿元，落实率为94.9%；落实公益性工程维修养护经费123.7亿元，落实率为79.9%。实行管养分离（包括内部管养分离）的水管单位9563个，占水管单位总数的79.1%。截至2018年年底，累计批准国家级水利风景区878个，其中：水库型373个，自然河湖型195个，城市河湖型195个，湿地型47个，灌区型31个，水土保持型37个。

农村水利改革。全年吸引社会资本投入高效节水灌溉和农村饮水安全巩固提升的资金，分别约占总投资的8.4%和16.8%。围绕三大类工程、13项改革措施，完成了全国100个县农田水利设施产权制度改革和创新运行管护机制试点任务。大型灌区、泵站等管理单位公益性人员基本支出和公益性工程维修养护经费落实率分别达到84%和35%。

水价改革。全年国家发展改革委、财政部、水利部、农业农村部联合印发《关于加大力度推进农业水价综合改革的通知》等文件,明确年度改革重点任务,将计划新增改革实施面积分解到各省份。截至2018年年底,农业水价综合改革实施面积累计超过1.6亿亩,其中2018年新增农业水价综合改革面积约1.1亿亩。实施农业水价综合改革的地区节水成效初显,工程管护水平显著提高。

水利规划和前期工作。2018年,中央层面审批水利规划14项,其中:国家发展改革委审批《洞庭湖水环境综合治理规划》1项,水利部审批13项。组织完成了《水利改革发展"十三五"规划》实施情况中期评估,加快推进第三次全国水资源调查评价,印发全国水资源调查评价水生态调查评价补充技术细则。积极推进全国层面重点水利规划工作,组织编制《防汛抗旱水利提升工程建设实施方案》《水利工程补短板总体方案》《全国重点区域水生态修复与治理总体方案》等;经报请国务院同意,会同有关部门联合印发《华北地区地下水超采综合治理行动方案》。加快推进重点流域和主要支流综合规划审批,扎实推进国家重大战略水利专项规划。围绕中央实施乡村振兴战略决策部署,积极推进乡村振兴战略水利规划工作取得初步成果。国家发展改革委批复项目11项,其中可行性研究报告10项、工程规划1项,总投资751.3亿元。水利部批复初步设计5项,总投资908.2亿元。

水土保持管理。2018年,全国共审批生产建设项目水土保持方案3.79万个,减少新增人为水土流失面积10586.33平方公里。全年完成生产建设项目的水土保持设施验收报备0.91万个,下达责令停止水土保持违法违规行为、限期补办手续、限期缴纳等文书1.36万个。

农村水电管理。截至2018年年底,19个省份累计创建了165座绿色小水电站。积极推进农村水电站安全生产标准化建设,全国累计建成2505座安全生产标准化电站,其中一级电站68座。福建、浙江、江西、河南、湖北、广东、重庆、四川、陕西、甘肃等10多个省

（直辖市）已出台水电站生态流量监督管理文件，对生态流量核定、泄放措施和监督管理作出明确规定。全国共有571条河流、1191个生态改造项目、1368个增效扩容项目完成改造，累计修复减脱水河段1000多公里。

水利移民。全年开工建设集中安置点760个，新建农村集中安置住房678.8万平方米。搬迁人口18.8万人，其中：农村移民搬迁18.3万人，城集镇移民搬迁0.5万人。生产安置23.6万人，其中：农业生产安置11.2万人，逐年补偿3.7万人，货币补偿安置（自行安置）6.9万人，养老保障安置1万人，投亲靠友安置0.1万人，其他安置0.7万人。国家核定上一年度新建大中型水库农村移民后期扶持人数6.9万人。

水利监督。全年水利行业共发生生产安全事故（事件）8起，死亡14人。全国省级及以上水行政主管部门组织10795个检查组赴现场开展水利安全巡查、检查和专项整治活动，排查治理隐患67896个。审定公布水利安全生产标准化一级单位82家，其中水利水电工程施工企业62家，水利工程项目法人7家，水利工程管理单位11家；完成水利部负责的水利水电工程施工企业相关责任人员安全生产考核10929余人。2018年水利部共派出6个批次106个稽查组对全国285个项目进行稽查，下发"一省一单"稽查整改意见104份。流域机构和省级水行政主管部门开展自主稽查，共派出稽查组458个，稽查项目1875个，下发整改通知790份。2018年水利部共派出9个批次385个督查组对全国 4702 座小型水库安全运行情况进行专项督查，下发"一省一单"通报 99 份。组织开展南水北调工程监督稽查 148 组，实现东、中线监管全覆盖，组织召开南水北调工程运行监督会商会和专项问题研判会 9 次，对严重问题实施责任追究 6 次，确保南水北调工程安全平稳运行。

依法行政。全年水行政法规、水利部规章和规范性文件未有修改、清理和废止发生。2018 年全国立案查处水事违法案件 23578 件，结案 19612 件，结案率 83.2%；各级水利部门共调处水事纠纷 27 件，解决 8 件；水利部共办结行政复议案件 45 件，办理行政应诉 12 件。

行政许可。全年水利部（包括部机关和各流域机构）共受理行政审批事项 1615 件，办结 1563 件。其中：水工程建设规划同意书审核 27 件，水利基建项目初步设计文件审批 5 件，取水许可发放 311 件（新审批 99 个、延续 154 个、变更 58 个），非防洪建设项目洪水影响评价报告审批 30 件，河道管理范围内建设项目工程建设方案审批 319 件，生产建设项目水土保持方案审批 62 件，国家基本水文测站设立和调整审批 5 件，国家基本水文测站上下游建设影响水文监测工程的审批 29 件，水利工程建设监理单位资质认定（新申请、增项、晋升、延续）521 件，水利工程质量检测单位甲级资质认定（新申请、增项、晋升、延续）250 件。

水利科技。全年国家立项安排3.95亿元资金用于水利科技项目，其中：组织承担国家重点研发计划"水资源高效开发利用"等涉水重点专项项目19项，水利技术示范项目84项。水利科技成果获国家科技进步二等奖4项，国家技术发明二等奖2项。截至2018年年底，水利系统共有国家和部级重点实验室12个，国家和部级工程技术研究中心15个。落实中央级科学事业单位修缮购置专项资金11985万元，落实中央财政公益性科研院所基本科研业务费10744万元。发布水利技术标准29项，在编126项，水利行业现行有效标准达856项。

国际合作。全年共签署水利国际合作协议5份，在华举办多边、双边高层圆桌会议或技术交流研讨会14次，亚洲开发银行、全球环境基金开展的4个项目进展顺利，中瑞、中丹、中法、中芬合作项目和国际科技合作项目稳步开展。

7 水利行业状况

水利单位。截至2018年年底,水利系统内外各类县级及以上独立核算的法人单位25602个,从业人员107.4万人。其中:机关单位2724个,从业人员12.6万人,比上年减少2.3%;事业单位17682个,从业人员58.8万人,比上年增加1.9%;企业4139个,从业人员34.4万人,比上年减少4.4%;社团及其他组织1057个,从业人员1.6万人,比上年增加166.7%。全国共有水利水电工程施工总承包特级资质企业27家,水利水电工程施工总承包一级资质企业263家。

职工与工资。截至2018年年底,全国水利系统从业人员90.3万人,比上年减少3.2%。其中:全国水利系统在岗职工87.9万人,比上年减少2.7%。在岗职工中,部直属单位在岗职工6.6万人,比上年增加3.4%,地方水利系统在岗职工81.3万人,比上年减少3.2%。全国水利系统在岗职工工资总额为802.7亿元,全国水利系统在岗职工年平均工资9.1万元。

职工与工资情况

	2008年	2009年	2010年	2011年	2012年	2013年	2014年	2015年	2016年	2017年	2018年
在岗职工人数/万人	105.6	103.7	106.6	102.5	103.4	100.5	97.1	94.7	92.5	90.4	87.9
其中：部直属单位/万人	7.2	7.2	7.4	7.5	7.4	7.0	6.7	6.6	6.4	6.4	6.6
地方水利系统/万人	98.4	96.5	96.3	95.0	96.0	93.5	90.4	88.1	86.1	84.0	81.3
在岗职工工资/亿元	234.4	264.7	297.9	351.4	389.1	415.3	451.4	529.4	640.5	739.1	802.7
年平均工资/(元/人)	22143	25633	28816	34283	37692	41453	46569	55870	69377	83534	91307

全国水利发展主要指标（2013—2018 年）

指标名称	单位	2013 年	2014 年	2015 年	2016 年	2017 年	2018 年
1. 灌溉面积	千公顷	69481	70652	72061	73177	73946	74542
2. 耕地灌溉面积	千公顷	63473	64540	65873	67141	67816	68272
其中：本年新增	千公顷	1552	1648	1798	1561	1070	828
3. 节水灌溉面积	千公顷	27109	29019	31060	32847	34319	36135
其中：高效节水灌溉面积	千公顷	14271	16114	17923	19405	20551	21903
4. 万亩以上灌区	处	7709	7709	7773	7806	7839	7881
其中：30 万亩以上	处	456	456	456	458	458	461
万亩以上灌区耕地灌溉面积	千公顷	30216	30256	32302	33045	33262	33324
其中：30 万亩以上	千公顷	11252	11251	17686	17765	17840	17799
5. 自来水普及率	%			76	79	80	81
农村集中供水率	%			82	84	85	86
6. 除涝面积	千公顷	21943	22369	22713	23067	23824	24262
7. 水土流失治理面积	万平方公里	106.9	111.6	115.5	120.4	125.8	131.5
其中：新增	万平方公里	5.3	5.5	5.4	5.6	5.9	6.4
8. 水库	座	97721	97735	97988	98460	98795	98822
其中：大型水库	座	687	697	707	720	732	736
中型水库	座	3774	3799	3844	3890	3934	3954
水库总库容	亿立方米	8298	8394	8581	8967	9035	8953
其中：大型水库	亿立方米	6529	6617	6812	7166	7210	7117
中型水库	亿立方米	1070	1075	1068	1096	1117	1126
9. 全年水利工程总供水量	亿立方米	6183	6095	6103	6040	6043	6016
10. 堤防长度	万公里	27.7	28.4	29.1	29.9	30.6	31.2
保护耕地	千公顷	42573	42794	40844	41087	40946	41351
堤防保护人口	万人	57138	58584	58608	59468	60557	62785
11. 水闸总计	座	98192	98686	103964	105283	103878	104403
其中：大型水闸	座	870	875	888	892	892	897

续表

指标名称	单位	2013年	2014年	2015年	2016年	2017年	2018年
12. 年末全国水电装机容量	万千瓦	28026	30183	31937	33153	34168	35226
全年发电量	亿千瓦时	9304	10661	11143	11815	11967	12329
13. 农村水电装机容量	万千瓦	7119	7322	7583	7791	7927	8044
全年发电量	亿千瓦时	2233	2281	2351	2682	2477	2346
14. 当年完成水利建设投资	亿元	3757.6	4083.1	5452.2	6099.6	7132.4	6602.6
按投资来源分：							
（1）中央政府投资	亿元	1729.8	1648.5	2231.2	1679.2	1757.1	1752.7
（2）地方政府投资	亿元	1542.0	1862.5	2554.6	2898.2	3578.2	3259.6
（3）国内贷款	亿元	172.7	299.6	338.6	879.6	925.8	752.5
（4）利用外资	亿元	8.6	4.3	7.6	7.0	8.0	4.9
（5）企业和私人投资	亿元	160.7	89.9	187.9	424.7	600.8	565.1
（6）债券	亿元	1.7	1.7	0.4	3.8	26.5	41.6
（7）其他投资	亿元	142.1	176.5	131.7	207.1	235.9	226.3
按投资用途分：							
（1）防洪工程	亿元	1335.8	1522.6	1930.3	2077.0	2438.8	2175.4
（2）水资源工程	亿元	1733.1	1852.2	2708.3	2585.2	2704.9	2550.0
（3）水土保持及生态建设	亿元	102.9	141.3	192.9	403.7	682.6	741.4
（4）水电工程	亿元	164.4	216.9	152.1	166.6	145.8	121.0
（5）行业能力建设	亿元	52.5	40.9	29.2	56.9	31.5	47.0
（6）前期工作	亿元	40.7	65.1	101.9	174.0	181.2	132.0
（7）其他	亿元	328.2	244.2	337.5	636.2	947.5	835.8

说明：1. 本公报不包括香港特别行政区、澳门特别行政区以及台湾省的数据。
   2. 水利发展主要指标分别于2012年、2013年与第一次全国水利普查数据进行了衔接。
   3. 农村水电的统计口径为单站装机容量5万千瓦及以下的水电站。

2018 STATISTIC BULLETIN ON CHINA WATER ACTIVITIES

Ministry of Water Resources, P. R. China

The year of 2018 marks the beginning of implementation of missions proposed by the 19th CPC National Congress and is also crucial for taking over from the past and setting a new course for water governance in the future. According to institutional reform plan of the CPC Central Committee, the Three Gorges Office of the State Council and Office of the South-to-North Water Transfer Project Construction Committee of the State Council were merged into the Ministry of Water Resources, so as to optimize the institutional functions and make administrative operations more efficient and effective, embarking on a new journey of water undertaking. The Party Central Committee and the State Council has paid great attention on water resources management. Under the guidance of Xi Jinping's thought of socialism with Chinese characteristics for a new era, the water departments at each level had achieved tremendous development in water field over the past year, through fulfillment of duties faithfully and energetically and respond to new challenges in an effective way.

I. Investment in Fixed Assets

Completed investment for water project construction in 2018 amounted to 660.26 billion Yuan, with a decrease of 52.98 billion Yuan, accounting for 7.4% comparing to the year of 2017, which includes 487.72 billion Yuan put into construction project with a 3.8% decrease; 28.09 billion Yuan for installation project with an increase of 5.7%; 21.44 billion Yuan for purchase of machinery, electric equipment and instruments, with an increase of 1.3%; and 123.01 billion Yuan for other purposes (including compensation of resettlement and land acquisition), with a decrease of 22.4%.

	2011 /billion Yuan	2012 /billion Yuan	2013 /billion Yuan	2014 /billion Yuan	2015 /billion Yuan	2016 /billion Yuan	2017 /billion Yuan	2018 /billion Yuan	increase /%
Total completed investment	308.60	396.42	375.76	408.31	545.22	609.96	713.24	660.26	−7.4
Construction project	210.32	273.65	278.28	308.64	415.08	442.20	506.97	487.72	−3.8
Installation project	12.17	23.78	17.36	18.50	22.88	25.45	26.58	28.09	5.7
Purchase of machinery, equipment and instruments	11.52	17.81	16.11	20.61	19.87	17.28	21.17	21.44	1.3
Others (including compensation of resettlement and land acquisition)	74.59	81.18	64.02	60.56	87.39	125.03	158.52	123.01	−22.4

In the total completed investment, 217.54 billion Yuan was allocated to the construction of flood control projects, 10.8% less than that in 2017; 255.00 billion Yuan for the construction of water resources projects, 5.7% less than that in 2017; 74.14 billion Yuan for soil and water conservation and ecological restoration, 8.6% increase than that in 2017; and 113.58 billion Yuan for special projects of hydropower development and capacity building, decreased by 13.0% comparing to the year before.

The competed investment for seven major river basins reached 510.86 billion Yuan, of which 149.4 billion Yuan was invested in river basins in the southeast, southwest and northwest of China. Completed investments in eastern, middle, western and northeast regions were 233.53 billion Yuan, 163.88 billion Yuan, 241.77 billion Yuan and 21.08 billion Yuan, accounting 35.4%, 24.8%, 36.6%, and 3.2% of the total, respectively.

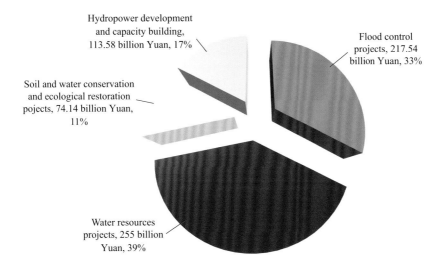

Completed investment of projects in 2018

Of this total competed investment, the Central Government contributed 11.66 billion Yuan, and local governments contributed 648.60 billion Yuan. The completed investment of large and medium-sized projects was 118.35 billion Yuan; the completed investment of small-sized and other projects was 541.91 billion Yuan; the completed investment of newly-constructed project was 514.96 billion Yuan; and the completed investment of reconstruction and expansion was 145.30 billion Yuan.

The newly-added-fixed assets for water conservancy construction totaled 361.08 billion Yuan in 2018. By the end of 2018, the accumulated completed investment of projects under construction was 1,674.59 billion Yuan, and the rate of completed investment reached 63.0%. The newly-added fixed assets totaled 945.92 billion Yuan and the rate of investment transferred into fixed assets was 56.5%, an increase of 1.0% to the year before.

A total of 27,930 water projects were under construction in 2018, with a total investment of 2,749.98 billion Yuan, an increase of 10.0% comparing to that of the year before. The projects with Central Government finance were 16,928 with an increase of 8.8% comparing to the year before. The total funds used by projects under construction reached 1,420.47 billion Yuan and increased 7.1% comparing to the year before. There were 19,786 newly-constructed projects in 2018 with an increase of 0.3%, and newly-added investment was 606.02 billion Yuan with a decrease of 33.2% to the year before. The completed civil works of earth, stone and concrete structures were 3.38 billion m^3, 450 million m^3, and 100 million m^3, respectively. By the end of 2018, the ratio of completed quantity of earthwork, stonework, and concrete of the under-construction projects were 78.2%, 67.7% and 63.4%, respectively.

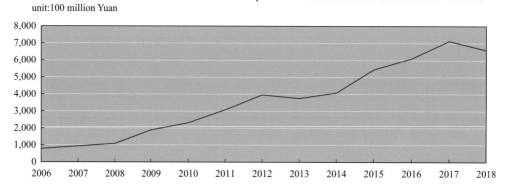

Completed investment for fixed water assets

II. Key Water Projects Construction

Harness of large rivers and lakes. In 2018, there were 5,406 river harness projects under construction, including 535 flood control dyke and embankment construction, 658 projects for large river and main tributary control, and 3,677 medium and small river control works, 536 flood diversion and storage areas or other projects. By the end of 2018, the accumulated investment in projects under construction was 338.63 billion Yuan, with a completion rate of 64.9%. River regime control and river course training and restoration had been undertaken in the middle and lower reaches of the Yangtze River. The recent flood control works in the lower reaches of the Yellow River have been quickened. Huai River improvement project has been accelerated, with 29 out of 38 projects under construction, among which, 6 projects were completed and started benefit generation. Control of three rivers project in the Northeast of China were nearly completed. There were 18 out of 21 projects for Comprehensive Improvement of Water Environment of Taihu Lake started construction, among which 12 projects completed for benefit generation.

Reservoir projects. There were 1,033 reservoir projects under construction in 2018. By the end of 2018, the completed investment of under-construction projects reached 294.26 billion Yuan, accounting for 58.1% of the total completed investment. Following large reservoirs started construction, namely Dongzhuang in Shaanxi, Nianpanshan in Hubei, Bailai in Fujian, Dongtaizi in Inner Mongolia, Niuling in Anhui, as well as 7 medium reservoirs, namely Qinjing and Yujian. There were 34 medium reservoir projects, including Shuangxi Reservoir and Huangsha Reservoir, have been accelerating construction. Damming of reservoir were completed for Huangzangsi in the Heihe River Basin, Huokou in Fujian Province, Huangjiawan in Guizhou Province, Chemabi in Yunnan Province and Duoying in Guangxi Autonomous Region. Houshan Reservoir in Liaoning Province was impounded.

Water allocation projects. The yearly investment for water allocation projects under construction reached to 732.78 billion Yuan. The completed investment had accumulated to 409.5 billion Yuan, accounting for 55.9% of the total. A total of 71 river-lake connecting systems were constructed, in order to improve connectivity of about 379 rivers (lakes or reservoirs).

Irrigation, drainage and rural water supply. The completed investment for strengthening and improving safe drinking water supply reached 57.36 billion Yuan, among which 7.63 billion Yuan from central government subsidy, with a beneficial population of 78 million of which 4.36 million listed in national plan for poverty reduction. By the end of 2018, the rural population access to tap water supply made up a percentage of 81% and the percentage of population with centralized water supply system raised to 86%. The Central Government allocated 13.756 billion Yuan for the construction of large and medium irrigation and drainage systems and rehabilitation of irrigation districts for water saving purpose. There were 22.5 billion Yuan allocated to the construction of highly-efficient water-saving farmland waterworks. The newly-added effective irrigated area reached 828,000 ha; new-added water-saving irrigated area was 2,022,000 ha and newly-added highly-efficient water-saving irrigated area was 1,536,000 ha.

Rural hydropower and electrification. In 2018, the completed investment of rural hydropower station construction amounted to 10.01 billion Yuan; the newly increased hydropower stations were 194, with a total installed capacity of 1.643 million kW, among which the newly increased installed capacity amounted to 1.161 million kW, and the increased installed capacity by rehabilitation accounts to 0.482 million kW.

Soil and water conservation. A total of 88.63 billion Yuan was allocated to construction of soil and water conservation and ecological restoration project in 2018, with an accumulated investment of 51.19 billion Yuan. The newly-added areas with soil conservation measures reached 64,000 km^2, of which the area under National Major Project for Soil Conservation was 12,500 km^2. There were 560 silt-retention dam on Loess Plateau at high risk were strengthened and rehabilitated.

Capacity building. The completed investment for capacity building in 2018 was 4.53 billion Yuan, of which 710 million Yuan spent on procurement of communication equipment for flood control, 1.31 billion Yuan for hydrological facilities, 190 million Yuan for scientific research and education facilities and 2.32 billion Yuan for others.

III. Key Water Facilities

Embankments and water gates. By the end of 2018, the completed river dykes and embankments ranging at Grade-V or above had a total length of 312,000 km❶. The accumulated length of dykes and embankments met the standard reached 218,000 km, with a percentage of 69.8% of the total, among which the Grade-I and Grade-II dykes and embankments up to the standard reached 34,000 km, with a reaching standard rate of 80.5%. These embankments can protect 630 million people and 41,000 ha of cultivated land. The number of water gates with a flow of 5 m^3/s increased to 104,403, of which 897 were large water gates. Divided by types of water gates, there were 8,373 flood diversion sluices, 18,355 drainage/return water sluices, 5,133 tidal barrages, 14,570 water diversion intakes and 57,972 controlling gates.

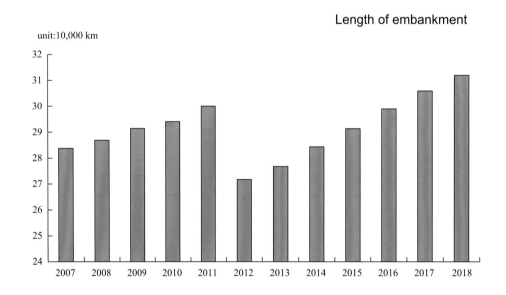

❶ The length of embankment before 2011 includes embankment below Grade-V.

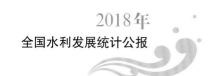

Reservoirs and water complexes. The number of reservoirs in China reached 98,822, with a total storage capacity of 895.3 billion m^3, of which 736 reservoirs belong to large reservoirs with a total capacity of 711.7 billion m^3, accounting 79.5% of the total; 3,954 medium-sized reservoirs with a total capacity of 112.6 billion m^3, accounting 12.6% of the total.

Tube wells and pumping stations. A total of 5.101 million tube wells, with a daily water abstraction capacity equal or larger than 20 m^3 or an inner diameter larger than 200 mm, were employed for water supply in the whole country. A total of 95,468 pumping stations that have an installed flow of 1 m^3/s or installed voltage above 50 kW were in operation, among which 376 categorized as larger pumping stations, 4,276 medium-size and 90,816 small-size pumping stations.

Irrigation systems. The irrigation districts with an area equal or above 2,000 mu added to 22,873, with a total effective irrigated area of 37.752 million ha, in which the irrigation districts equal or above 500,000 mu reached 175, with a total irrigated area of 12.399 million ha; the irrigation districts covering an area from 300,000 – 500,000 mu was 286, with a total irrigated area of 5.4 million ha. By the end of 2018, the total irrigated area and irrigated area of cultivated land reached to 74.542 million ha and 68.272 million ha respectively, taking 50.7% of the total cultivated land in China. The areas with water-saving irrigation facilities totaled 36.135 million ha, among which 11.338 million ha equipped with sprinkler or micro irrigation systems and 10.566 million ha installed low-pressure pipes.

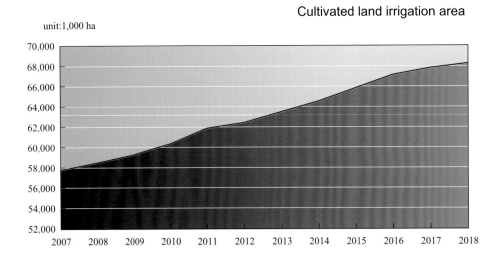

Cultivated land irrigation area

Rural hydropower and electrification. By the end of 2018, hydropower stations built in rural areas totaled 46,515, with an installed capacity of 80.435 million kW, accounting for 22.8% of the national total. The annual power generation by these hydropower stations reached to 234.56 billion kW·h, accounting for 19.0% of the national total.

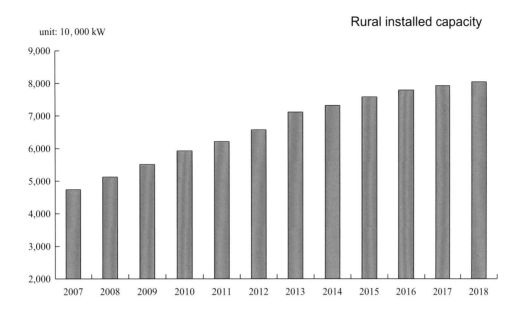

Soil and water conservation. By the end of 2018, the restored eroded areas reached 1.315 million km² [1]; and the forbidden area for ecological restoration accumulated to 234,300 km². Dynamic monitoring for soil and water loss had been conducted by central and local authorities in a collaborated manner, so as to cover the whole area of the country. With a total area of about 9.6 million km² being monitored, dynamic changes of soil and water loss in county-level administrative areas and key areas were well understood.

---

[1] Statistical data in 2012 is integrated with the data of first national census for water.

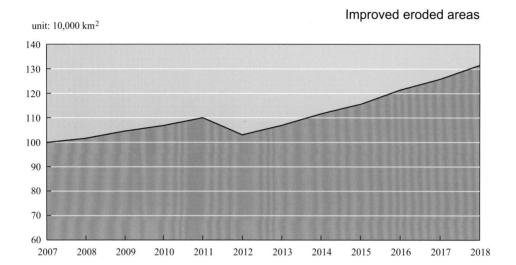

Hydrological station networks. In 2018, the number of hydrological stations of all kinds was totaled 121,097 in the whole country, including 3,154 national basic hydrologic stations, 4,099 special hydrologic stations, 13,625 gauging stations, 55,413 precipitation stations, 19 evaporation stations, 26,550 soil moisture monitoring stations, 14,286 water quality stations, 3,908 groundwater monitoring stations and 43 experimental stations. There were 66,439 various kinds of hydrological monitoring stations that provide hydrological information to flood control commanding headquarters at and above the county level; 1,887 various kinds of hydrological monitoring stations for early warning and forecasting. A total of 1,616 hydrological monitoring stations were equipped for online flow measurement. A water quality monitoring system, including 332 monitoring centers at central, basin, provincial and local levels, had been formed and spread over major rivers, lakes, reservoirs and aquifers.

Water networks and information systems. By the end of 2018, water resources departments at and above the provincial level equipped 5,764 sets of varied kinds of servers, forming the storage capacity of 18PB, with a total of 2.4PB storage of various kinds of data and information. The water resources departments at and above the county levels had installed 3,895 sets of various kinds of satellite equipment, 6,549 flood forecasting stations for short message transmission from Beidou Satellite, 48 vehicles for emergency communication, 819 Unmanned Aerial Vehicle (UAV) and 5,313 cluster communication terminals in operation. In terms of data collection and video monitoring, a total of 450,000 gathering points were available for water departments at and above the provincial level to receive various kinds of water-related information, among which 227,600 points were used for collecting data of hydrology, water resources and soil and water conservation. There were 225,000 gathering points for monitoring safety of large and medium reservoirs.

IV. Water Resources Utilization and Protection

Water resources conditions. The total national water resources in 2018 was 2,746.25 billion m^3, remained almost the same as the normal years. Mean annual precipitation was 682.5 mm that was 6.2% more than the normal years and 2.7% more than the year before. By the end of 2018, total storage of 669 large and 3,602 medium-size reservoirs were 410.43 billion m^3, 3.8 billion m^3 less than that in early 2018.

Water resources development. In 2018, the newly-increased water supply capacity of water projects above designated size❶ reached 10.16 billion m^3. By the end of 2018, the total water supply capacity in China reached 867.68 billion m^3, including 56.71 billion m^3 of water supplied by utilities at county level, 232.37 billion m^3 by reservoirs, 210.51 billion m^3 by river-lake diversion schemes, 175.47 billion m^3 by river-lake pumping stations, 140.09 billion m^3 by tube wells, 35.79 billion m^3 by ponds, weirs and cellars, and 16.84 billion m^3 by unconventional water sources.

Water resources utilization. In 2018, the total water supply amounted to 601.55 billion m^3, among which 82.3% came from surface water, 16.2% from underground water and 1.5% from other water sources. The total water consumption amounted to 601.55 billion m^3, among which domestic water use

---

❶ Water projects above designated size include: reservoirs with a total storage capacity greater than or equal to 100,000 m^3, river and lake water intake pumping stations with an installed flow greater than or equal to 1 m^3/s or with an installed capacity greater than or equal to 50 kW, river and lake water intake sluices with water discharge through sluice greater than or equal to 1 m^3/s, irrigation tube wells with a wellhead & sidewall pipe inner diameter greater than or equal to 200 mm, and tube wells with a daily supply capacity greater than or equal to 20 m^3.

amounted to 85.99 billion m^3 or 14.3% of the total; industrial water use 126.16 billion m^3 or 21.0% of the total; agricultural water use 369.31 billion m^3 or 61.4% of the total; artificial recharge for environmental and ecological use 20.09 billion m^3 or 3.3% of the total. Comparing to that of the year before, water consumption decreased by 2.79 billion m^3, in which agricultural water use decreased by 7.33 billion m^3, industrial water use decreased by 1.54 billion m^3, and domestic water use and artificial recharge for environmental and ecological use increased by 2.18 billion m^3 and 3.9 billion m^3 respectively. Water consumption per capita in 2018 was 432 m^3 in average. The coefficient of effective irrigated water use was 0.554. Water use of 10,000 Yuan GDP (at comparable price of the same year) was 66.8 m^3. Water use of industrial production value added per 10,000 Yuan (at comparable price of the same year) was 41.3 m^3. At the comparable price of the same year, water use of 10,000 Yuan GDP and water use of industrial production value added per 10,000 Yuan decreased by 6.6% and 6.9% comparing to 2017.

Water quality of rivers and lakes. According to water quality assessment on 262,000 km long of rivers, 81.6% of the river met the class-I to class-III water quality standard.

V. Flood Control and Drought Relief

In 2018, the overall damage caused by flood and water-logging disasters was relatively less than other years, and the direct economic loss of flood disaster accounted for 0.18% of GDP in the year. A total of 6.42 million ha of cultivated land were affected by floods, 3.13 million ha of farmland had no harvest, 56 million people affected, 187 people dead, and 32 people missing. A total of 85.1 million houses were destroyed and 83 cities suffered from inundation. The disasters resulted in 161.547 billion Yuan of direct economic losses, among which the direct losses with water infrastructures reached 25.798 billion Yuan. Provinces suffered from severe flooding include Sichuan, Shandong, Guangdong, Gansu, Inner

Mongolia, and Yunnan. Death toll or people missing caused by mountain flood took 73.52% of the total. The direct economic loss caused by typhoons took 41.71% of the total loss as a result of flood and waterlogging disasters.

In 2018, no large scale drought occurred in the whole country. The seriously affected areas include provinces or autonomous region of Inner Mongolia and Liaoning. The affected farmland was 7.38 million ha and areas with no harvest reached 3.67 million ha, with a direct economic losses of 48.362 billion Yuan. A total of 3.07 million urban and rural population and 4.623 million man-feed big animals and livestock suffered from temporary drinking water shortage.

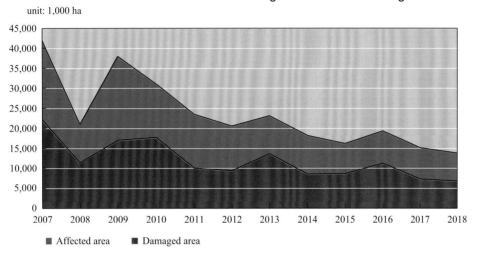

In 2018, the funds allocated to defense extraordinary floods and droughts amounted to 3.506 billion Yuan, among which 3.099 million Yuan for extraordinary floods and 0.406 billion Yuan for extraordinary droughts. Thanks to the efforts of flood control, 3.18 million ha of cultivated land were prevented from inundation and 75 times of flooding in cities were avoided, resulting in economic benefits amounting to 23.998 billion Yuan. Drinking water was provided to 4.77 million people in rural and urban areas as well as 2.44 million big animals and livestock in order to alleviate temporary water shortage. The area with anti-drought measures reached 23.42 million ha that prevented a loss of 22.138 billion kg of grain. The accumulative inputs for flood disaster relief include 4.68 million person-time, 17,800 vessel-time, 200,000 shifts of transportation and 146,800 shifts of mechanical equipment. The consumed materials valued 1.315 billion Yuan. The person-time of emergency evacuation was 8.07 million. There were 4.68 million tube wells, 36,600 pumping stations, 5.69 million mobile devices and 936,500 vehicle-time of various kinds of water transporting vehicle employed for drought relief.

VI. Water Management and Reform

River (Lake) chief system. By the end of 2018, river (lake) chief system had been fully implemented across the country. More than 300,000 river (lake) chiefs were named at four levels of province, city, county and township. More than 900,000 river (lake) chiefs including river inspectors and river guards were assigned at the village level. All provinces, autonomous regions or municipalities set up river chief offices at provincial, municipal, county levels, and counterpart management systems with a working pattern of taking responsibilities by the government leaders, taking the lead by water resources departments, coordinated by other governmental departments and participated by the whole society.

Most stringent water resources management. The Ministry of Water Resources, in collaboration with the National Development and Reform Commission, completed performance evaluation of implementing most stringent water resources management system in 31 provinces and autonomous regions in 2017. Seven provinces (municipalities directly under the Central Government) of Jiangsu, Shandong, Anhui, Chongqing, Beijing, Zhejiang and Shanghai were rated as excellent. Water allocation plans of 13 trans-provincial river basins, including Taihu Lake and Songhua River, were approved. Water allocation was completed for river basins across border of city or county in 13 provinces and autonomous regions, including Hebei and Liaoning. In 2017-2018 of water diverted year, the Phase-I of Eastern Route of South-to-North Water Diversion diverted 1.088 billion m^3 water to Shandong; the Phase-I of Middle Route of South-to-North Water Diversion diverted 7.458 billion m^3 water to Beijing, Tianjin, Hebei and Henan provinces. As a result, water security in the receiving areas has been greatly enhanced. Water resources has been regulated in an integrated way in major river basins, which ensured continuous flow at the mainstream of the Yellow River for 19 consecutive years, and no dry up of the East Dongjuyan Lake in the lower reaches of the Heihe River for 14 years. Evaluation Criteria of Water-saving Society was issued that formed a base for overall assessment of water efficiency increase by using county as a unit. There were 65 counties (districts) in 8 provinces or autonomous regions or municipalities such as Beijing passed the final evaluation. The pilot project of 105 water ecological civilization cities has made smooth progress, among which 96 pilots passed check and acceptances. Pilot projects of groundwater recharge in river and lake areas were conducted, with a total recharge of 627 million m^3 water into three rivers in Hebei Province, by using diverted water from the Middle Route of South-to-North Water Diversion and reservoirs in Hebei Province. Phase-I of Eastern Route Scheme of South-to-North Water Diversion has deducted 1.9 billion m^3 for groundwater withdrawal in the water receiving area. Seven pilot projects for water right trading in Ningxia, Inner Mongolia, Guangdong, Henan, Gansu, Jiangxi, and Hubei were completed and passed check and acceptance. In 2018, China Water Exchange completed 51 entitlement trading with an amount of 1.331 billion m^3, with a total of 92 entitlements and trading of 2.774 billion m^3 accumulatively since it went into operation.

Operation and management. By the end of 2018, there were 14,325 water utilities included in the list for water management system reform, and adjusted and streamlined to 12,908 water utilities through cancel and merger, with a reduction of 10% comparing to the number of organizations before the reform. A total of 30.14 billion Yuan had been allocated for covering cost of managerial staff and operation and maintenance, which covered 88.1% of the total, among which, 17.77 billion Yuan were allocated to cover the staff expenses of public service, which accounted 94.9% of the total; 12.37 billion Yuan allocated to cover operation and maintenance of public service facilities, which accounted 79.9% of the total. Reform of management system by separating functions of administration with operation had been implemented in 9,563 utilities, accounted for 79.1% of the total. By the end of 2018, the approved national water scenery spots reached 878, including 373 reservoirs, 195 natural rivers and lakes, 195 lake or riverine cities, 47 wetlands, 31 irrigation districts and 37 soil conservation areas.

Reform in rural water resources management. In 2018, the funding from private sector to high-efficient water-saving irrigation and provision of safe drinking water amounted to 8.4% and 16.8% of the total investments respectively. Focusing on three types of projects and 13 reform measures, the pilot projects in 100 counties that implemented reform of property right of farmland waterworks and innovative operation and maintenance mechanism had completed. The coverage rates of cost of basic personal expenses in large irrigation districts and pumping station as well as operation and maintenance of public-good waterworks were 84% and 35% respectively.

Water pricing reform. In 2018, *Notice on Vigorously Promoting Comprehensive Agricultural Water Pricing Reform* was jointly issued by NSDC, Ministry of Finance (MOF), MWR, and Ministry of Agriculture and Rural Affairs. It clearly specifies the yearly tasks of reform, which had been allocated to each province based on newly-added areas defined by the plan. By the end of 2018, the area carrying out water pricing reform reached 160 million mu, among which about 110 million mu were newly added area in 2018. Great achievements had been made in water conservation thanks to the implementation of reform, which greatly improved operation and maintenance of waterworks in these regions.

Water resources planning and early-stage work. In 2018, there were 14 water resources plans approved by central government agencies, among which *the Comprehensive Plan of Water Environment Restoration in Dongting Lake* was approved by NDRC; 13 plans approved by MWR. The mid-term evaluation of implementation of 13*th Five-Year Plan for Water Resources Reform and Development* was completed. The third national water resources survey and assessment was accelerated. Supplementary technical guidelines on water ecology investigation and evaluation for national water resources survey and assessment was released. Steady progress had been made by major water project planning at the national level. *Implementation Plan for Construction Upgrading of Flood Control and Drought Relief Projects*, *Overall Plan for making up shortfall of water works and General Plan for Water Ecology Recovery and Restoration in National Key Regions* was compiled and approved by the State Council. *Action Plan for Comprehensive Control of Groundwater Overexploitation in North China* was jointly issued by MWR and other relevant departments. Comprehensive planning for major river basins and tributaries has been accelerated. Water plans in key national strategies has been proceeded. Preliminary results have been obtained in the advancement of strategic plan of water for rural revitalization, according to the decisions made by the Chinese government. In 2018, a total of 11 projects subject to NDRC approval were finalized, including 10 feasibility study reports and 1 project plan, with a total investment of 75.13 billion Yuan. There were 5 preliminary designs approved by MWR, with a total investment of 90.822 billion Yuan.

Soil and water conservation. In 2018, a total of 37,900 soil and water conservation plans of construction projects being examined and approved, covering a scope of 10,586.33 km^2 for protection and control. A total of 9,100 soil and water conservation projects completed check and acceptance. There were 13,600 official notices were given to order the cessation of illegal activities, and issue dates and deadlines for going through necessary procedures or making payment within a time limit.

Reform of hydropower management system. By the end of 2018, there were 165 small hydropower projects in 19 provinces awarded the title of green small hydropower station. Standardization for safety production of rural hydropower stations had been advocated and 2,505 hydropower stations of the kind have been completed, including 68 Class-I hydropower stations. Regulatory documents on ecological flow monitoring and supervision were issued for managing hydropower stations in more than 10 provinces and cities, including Fujian, Zhejiang, Jiangxi, Henan, Hubei, Guangdong, Chongqing, Sichuan, Shaanxi and Gansu. It stipulates the methods of verifying ecological flow, water release as well as monitoring and supervision. Rehabilitation was completed for 571 rivers, with 1,191 ecological restoration projects and 1,368 capacity expansion projects. The recovered river sections with low water level or drying-up accumulated to more than 1,000 km.

Water conservancy resettlement. There were 760 concentrated relocation sites constructed and centralized newly-constructed housing of 6.788 million km^2 in 2018. The resettled population was 188,000, among which 183,000 were relocated rural residents and 5,000 relocated urban residents. A total of 236,000 resettled people were arranged for production activities, among which 112,000 arranged for agricultural production, 37,000 people compensated on a yearly basis, 69,000 compensated by monetary means (arranged by themselves), 10,000 people

joined pension plans, 1,000 people seeking help from their relatives and friends and 7,000 people with other arrangements. According to approved estimation of the central government, the resettled people in rural areas, who received government support due to the newly-built large and medium reservoirs, reached 69,000 in the previous year.

Safety supervision. In 2018, there were 8 production accidents with 14 people dead. MWR, river basin authorities and water resources departments at provincial level organized 10,795 inspection teams for on-site inspection and overall investigation of safe production situation, and also inspections during flood season and highly-hazard chemicals as well as special inspections on electric fire risks. A total of 67,896 potential hazards were properly handled. It was announced that 82 enterprises approved to meet the level-1 evaluation standard for water safety production, including 62 water and hydropower constructing companies, 7 legal persons of water projects and 11 water project management units. Evaluation of about 10,929 project leaders and peoples in charge of safe production management had been taken. MWR sent 6 batches including 106 missions for inspection of 285 projects in 2018. There were 104 feedbacks of inspection and rectification comments that were delivered one by one to each province regarding prominent problems. Self-inspection of provincial water administrative departments had been encouraged and a total of 458 inspection groups were sent out for inspection of 1,875 projects, with 790 rectification announcements issued. In 2018, the ministry sent out 385 missions in 9 groups to inspect 4,702 small reservoirs for safety operation. The ministry circulated 99 correction notices that delivered to each province one by one. There were 148 inspection teams were dispatched to inspect South-to-North Water Project, realizing the full coverage of Eastern and Middle routes. There were 9 consultation conferences or special meetings were held for consultation and decision making, regarding issues found out during inspection of South-to-North Water Diversion Project. Accountability investigations were made on 6 cases of serious breach of duty, so as to ensure safe and stable operation of

South-to-North Water Diversion Project.

Legislation and administrative law enforcement. In 2018, there was no revision cleaned up or abolishment to water-related administrative regulations, ministerial norms and standards and normative documents. In 2018, the investigated illegal cases totaled 23,578 and 19,612 or 83.2% resolved. A total of 27 water disputes were mediated and 8 resolved at all levels. There were 45 administrative review cases and 12 administrative responses settled by the Ministry of Water Resources.

Administrative permits. There were 1,615 applications accepted by MWR and 1,563 water-related administrative approvals or permits authorized or extended, including 27 project plan approvals, 5 preliminary design reports of water construction projects, 311 water abstraction licenses (99 approvals for new application, 154 for renewal and 58 for modification), 30 evaluation reports of flood impact by non-flood control project, 319 plans of construction projects within the jurisdiction of river courses, 62 approvals of soil and water conservation plan of production and construction projects, 5 approvals for establishment and adjustment of National Basic Hydrological Stations, 29 approvals of hydrological monitoring projects for impact of construction at upper and lower of National Basic Hydrological Stations, 521 qualification approvals (including new application, extension, adding of new items or promotion) for construction supervisors of water resources projects; 250 Class-A qualification identifications (including new application, extension, adding of new items or promotion) for quality supervisors of water-related projects.

Water science and technology. A total of 395 million Yuan had been allocated to science and technology projects, including 19 special-subject and water-related scientific research projects listed in the National Key Research and Development Plan-Effective Development and Utilization of Water Resources, and 84 demonstration projects for water technologies. There were 4 water technological achievements won the National Sci-Tech Advance Award (second prize) and 2 projects won the second prize of National Technological Innovation Award. By the end of 2018, the numbers of national level or ministerial level labs were 12, and technical research centers were 15. Special funds for procurement and repairing of equipment of national scientific institutions amounted to 119.85 million Yuan. A total of 107.44 million Yuan had been allocated from central government finance as operation expenses for basic scientific studies of public research institutes. There were 29 water-related technical standards were made public and 126 standards were under drafting. The effective water-related technical standard reached 856 in total.

International cooperation. In 2018, a total of 5 water-related international cooperation agreements were signed. There were 14 multilateral and bilateral high-level round-table meetings and technical exchange symposiums or seminars held. There were 4 foreign funded projects financed by the Asian Development Bank and Global Environment Fund under smooth implementation. Bilateral cooperation project between China and Switzerland, Denmark, France and Finland as well as international science and technology cooperation projects were steadily progressed.

VII. Current Status of Water Sector

Water-related institutions. By the end of 2018, the legal entities with separate accounts that had engaged in water activities within the administrative jurisdiction at county level or above were totaled 25,602 that had 1.074 million employees. Among which governmental organizations was 2,724 with 126,000 employees, decreased by 2.3% than last year; public organizations 17,682 with 588,000 employees, increased by 1.9% than last year; 4,139 enterprises with 344,000 employees, decreased by 4.4% than last year; 1,057 societies and other institutions with 16,000 employees, increased by 166.7% than last year. There were 27 general construction contractors awarded highest qualification for water resources and hydropower project construction; 263 general construction contractors awarded grade-I qualification.

Employees and salaries. By the end of 2018, the employees of water sector were totaled 903,000, 3.2% lower than that the year before, of which in-service staff amounted to 879,000, 2.7% lower than last year. In the in-service staff, the staff working in the agencies directly under the Ministry of Water Resources was 66,000, 3.4% higher than last year; the staff working in local agencies was 813,000, 3.2% lower than last year. The total salary of nationwide in-service staff was 80.27 billion Yuan, and the annual average salary per in-service staff was 91,000 Yuan.

Employees and Salaries

	2008	2009	2010	2011	2012	2013	2014	2015	2016	2017	2018
Number of in service staff /10^4 persons	105.6	103.7	106.6	102.5	103.4	100.5	97.1	94.7	92.5	90.4	87.9
Of which: staff of MWR and agencies under MWR/10^4 persons	7.2	7.2	7.4	7.5	7.4	7.0	6.7	6.6	6.4	6.4	6.6
Local agencies/10^4 persons	98.4	96.5	96.3	95.0	96.0	93.5	90.4	88.1	86.1	84.0	81.3
Salary of in-service staff/10^8 Yuan	234.4	264.7	297.9	351.4	389.1	415.3	451.4	529.4	640.5	739.1	802.7
Average salary /(Yuan/person)	22,143	25,633	28,816	34,283	37,692	41,453	46,569	55,870	69,377	83,534	91,307

Main Index of National Water Resources Development (2013–2018)

Indicators	unit	2013	2014	2015	2016	2017	2018
1. Irrigated area	10^3 ha	69,481	70,652	72,061	73,177	73,946	74,542
2. Farmland irrigated area	10^3 ha	63,473	64,540	65,873	67,141	67,816	68,272
Newly-increased in 2018	10^3 ha	1,552	1,648	1,798	1,561	1,070	828
3. Water-saving irrigated area	10^3 ha	27,109	29,019	31,060	32,847	34,319	36,135
Highly-efficient water-saving irrigated area	10^3 ha	14,271	16,114	17,923	19,405	20,551	21,903
4. Irrigation districts over 10,000 mu	Unit	7,709	7,709	7,773	7,806	7,839	7,881
Irrigation districts over 300,000 mu	Unit	456	456	456	458	458	461
Farmland irrigated areas in irrigation districts over 10,000 mu	10^3 ha	30,216	30,256	32,302	33,045	33,262	33,324
Farmland irrigated areas in irrigation areas over 300,000 mu	10^3 ha	11,252	11,251	17,686	17,765	17,840	17,799
5. Rural population accessible to safe drinking water	%			76	79	80	81
Centralized water supply system	%			82	84	85	86
6. Flooded or waterlogging area under control	10^3 ha	21,943	22,369	22,713	23,067	23,824	24,262
7. Controlled or improved eroded area	10^4 km^2	106.9	111.6	115.5	120.4	125.8	131.5
Newly-increased	10^4 km^2	5.3	5.5	5.4	5.6	5.9	6.4
8. Reservoirs	Unit	97,721	97,735	97,988	98,460	98,795	98,822
Large-sized	Unit	687	697	707	720	732	736
Medium-sized	Unit	3,774	3,799	3,844	3,890	3,934	3,954
Total storage capacity	10^8 m^3	8,298	8,394	8,581	8,967	9,035	8,953
Large-sized	10^8 m^3	6,529	6,617	6,812	7,166	7,210	7,117
Medium-sized	10^8 m^3	1,070	1,075	1,068	1,096	1,117	1,126

Continued

Indicators	unit	2013	2014	2015	2016	2017	2018
9. Total water supply capacity of water projects in a year	10^8 m³	6,183	6,095	6,103	6,040	6,043	6,016
10. Length of dikes and embankments	10^4 km	27.7	28.4	29.1	29.9	30.6	31.2
Cultivated land under protection	10^3 ha	42,573	42,794	40,844	41,087	40,946	41,351
Population under protection	10^4 people	57,138	58,584	58,608	59,468	60,557	62,785
11. Total water gates	Unit	98,192	98,686	103,964	105,283	103,878	104,403
Large-sized	Unit	870	875	888	892	892	897
12. Total installed capacity by the end of the year	10^4 kW	28,026	30,183	31,937	33,153	34,168	35,226
Yearly power generation	10^8 kWh	9,304	10,661	11,143	11,815	11,967	12,329
13. Installed capacity of rural hydropower by the end of the year	10^4 kW	7,119	7,322	7,583	7,791	7,927	8,044
Yearly power generation	10^8 kWh	2,233	2,281	2,351	2,682	2,477	2,346
14. Completed investment of water projects	10^8 Yuan	3,757.6	4,083.1	5,452.2	6,099.6	7,132.4	6,602.6
Divided by different sources							
(1) Central government investment	10^8 Yuan	1,729.8	1,648.5	2,231.2	1,679.2	1,757.1	1,752.7
(2) Local government investment	10^8 Yuan	1,542.0	1,862.5	2,554.6	2,898.2	3,578.2	3,259.6
(3) Domestic loan	10^8 Yuan	172.7	299.6	338.6	879.6	925.8	752.5
(4) Foreign funds	10^8 Yuan	8.6	4.3	7.6	7.0	8.0	4.9
(5) Enterprises and private investment	10^8 Yuan	160.7	89.9	187.9	424.7	600.8	565.1
(6) Bonds	10^8 Yuan	1.7	1.7	0.4	3.8	26.5	41.6
(7) Other sources	10^8 Yuan	142.1	176.5	131.7	207.1	235.9	226.3
Divided by different purposes:							
(1) Flood control	10^8 Yuan	1,335.8	1,522.6	1,930.3	2,077.0	2,438.8	2,175.4
(2) Water resources	10^8 Yuan	1,733.1	1,852.2	2,708.3	2,585.2	2,704.9	2,550.0
(3) Soil and water conservation and ecological recovery	10^8 Yuan	102.9	141.3	192.9	403.7	682.6	741.4

Continued

Indicators	unit	2013	2014	2015	2016	2017	2018
(4) Hydropower	10^8 Yuan	164.4	216.9	152.1	166.6	145.8	121.0
(5) Capacity building	10^8 Yuan	52.5	40.9	29.2	56.9	31.5	47.0
(6) Early-stage work	10^8 Yuan	40.7	65.1	101.9	174.0	181.2	132.0
(7) Others	10^8 Yuan	328.2	244.2	337.5	636.2	947.5	835.8

Notes:

1. The data in this bulletin do not include those of Hong Kong, Macao and Taiwan.

2. Key indicators for water development and statistical data in 2012 and in 2013 is also integrated with the data of first national census for water.

3. Statistics of rural hydropower refer to the hydropower stations with an installed capacity of 50,000 kW or lower than 50,000 kW.

《2018年全国水利发展统计公报》编辑委员会

主　　　　任：叶建春
副　主　　任：石春先
委　　　　员：（以姓氏笔画为序）
　　　　　　　王　翔　王　静　付　涛　匡尚富　邢援越　朱闽丰
　　　　　　　朱　涛　任骁军　刘六宴　李原园　李晓静　张严明
　　　　　　　张清勇　陈茂山　姜成山　钱　峰　倪文进　倪　莉
　　　　　　　徐　洪　郭孟卓　郭索彦　曹纪文　曹淑敏　寇全安
　　　　　　　谢义彬　蔡建元

《2018年全国水利发展统计公报》主要编辑人员

主　　　　编：石春先
副　主　　编：谢义彬　吴　强
执　行　编　辑：汪习文　张光锦　乔根平
主要参编人员：（以姓氏笔画为序）
　　　　　　　万玉倩　王　为　王　伟　王位鑫　王　超　尤　伟
　　　　　　　曲　鹏　吕　烨　乔根平　刘永攀　刘宝勤　齐兵强
　　　　　　　杜崇玲　李春明　李笑一　李　益　吴梦莹　汪习文
　　　　　　　沈东亮　张光锦　张　岚　张岳峰　张晓兰　周　玉
　　　　　　　戚　波　喜　洋　蓝希龙
主要数据处理人员：（以姓氏笔画为序）
　　　　　　　王小娜　王明军　王鹏悦　刘　品　李　聪　郭　悦
　　　　　　　潘利业

◎ 主编单位
水利部规划计划司

◎ 协编单位
水利发展研究中心

◎ 参编单位
水利部办公厅
水利部政策法规司
水利部财务司
水利部人事司
水利部水资源管理司
全国节约用水办公室
水利部水利工程建设司
水利部运行管理司
水利部河湖管理司
水利部水土保持司
水利部农村水利水电司

水利部水库移民司
水利部监督司
水利部水旱灾害防御司
水利部水文司
水利部三峡工程管理司
水利部南水北调工程管理司
水利部调水管理司
水利部国际合作与科技司
水利部综合事业局
水利部信息中心
水利部水利水电规划设计总院
中国水利水电科学研究院